This book series is dedicated to changing the face of Science, Technology, Engineering and Math  one story at a time.  Dedicated to Dayten and Gabriel, who will live out my legacy. Thank you Eric, Malik, and Denise for your support and living in my legacy. Thank you Christian for pushing me.

-K. Renee

This book is dedicated to my 6th grade students.

-Theresa

Author:  K Renee Horton, Ph.D

Artist: Patrick Giles

It was the weekend, but Dr. H had no special plans. While most people she knew were going camping, Dr. H wanted to do something a bit more exciting. She hurried outside and waved to her purple buddy, Bouchet Beetle, who smiled back.

"Hi Dr. H, did you want to go on another space adventure?" BB asked.

"I sure do! We still have lots of planets to visit," Dr. H exclaimed, climbing into the driver's seat.

As soon as she closed the door, BB zoomed up into space. It was fun to be surrounded by billions of stars as they flew through space. Their last space trip stopped on Mars, so they traveled to the next planet in line, which was Jupiter. Jupiter was the largest planet, but spun very quickly.

"Did you know that one day on Jupiter only lasts about 10 hours?" Dr. H asked as BB gasped with shock.

"Are you serious? That's so short."

Jupiter was easy to spot in space, because it rotated so quickly and created colorful bands of gas. They landed on the surface to find it riddled with storms! BB tried to fly off the surface, but he was too heavy.

"I almost forgot! The gravity on Jupiter is much stronger, so we weigh two and a half times heavier," Dr. H explained, just as a storm swooped them into the air.

They flew around in a circle as the storm raged across Jupiter. BB had a fun idea as he hopped out of the storm and right into another. It was like hopscotch with storms!

"You have perfect timing, BB," Dr. H complimented her friend.

The pair giggled as they continued hopping from storm to storm.

With a final leap, he jumped right into a violent, red hued storm.

"This must be the big red spot. It's Jupiter's most violent storm, and has been going for over three hundred years," Dr. H exclaimed as they whooshed around the storm. "We could use this storm to propel us back into space."

BB nodded as he flew to the edge of the storm. They traveled higher and higher in the storm until BB flew straight into the air. They carried on through space to visit the next planet, which was Saturn. Before they reached the planet, they were amazed at Saturn's beautiful rings.

So, BB was a little surprised to uncover that the rings were made of up lots of rocks that rotated around the planet! They maneuvered between the zooming rocks and landed safely on the surface of Saturn.

Dr. H hopped out of BB as the pair explored the planet. Saturn was the second biggest planet in the Milky Way, but it looked much larger because of its rings. They remained perfectly still as they watched the stars move around them faster than clouds moved on Earth.

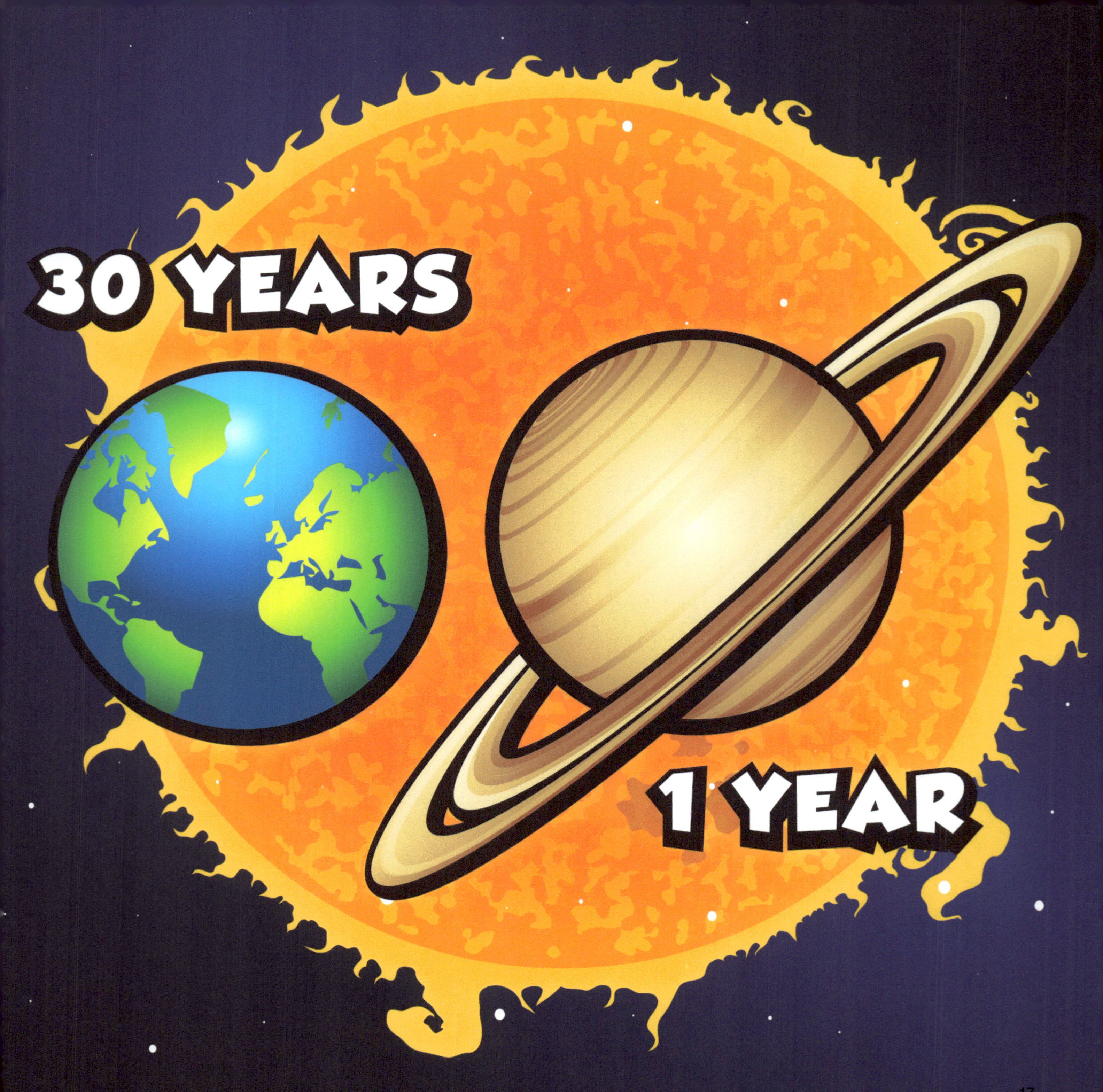

30 YEARS
1 YEAR

"Why are the stars moving so fast?" BB asked.
"We're the ones that are rotating fast, explained
Dr. H. "Instead of taking 24 hours to rotate fully like Earth
does, Saturn finishes rotating in 10 and a half  hours! A
year here is about 30 years back home."

When they were ready to leave, Dr. H hopped inside as BB came up with a plan to fly away. Since they weighed much more on Saturn, BB drove at top speed before take off. He continued to speed on until they were far enough from the planet to return to their normal weight.

They traveled through space until they reached the next planet, which was Uranus. Uranus also had rings, but the planet was much smaller. But as they flew closer, they noticed something odd about the planet. Uranus was the only one that spun on its side!

They landed on the surface to find it quite dark. It looked like it was night time on Uranus.

"I forgot an Earth year on Uranus is broken into 42 years of daylight and 42 in darkness. We must have arrived during the night," Dr. H stated.

Dr. H's whole body shivered as she grabbed a thick coat from BB's trunk, and slid it over her astronaut suit. Since the planet was dark and so far away from the sun, it was extremely chilly on the surface.

"Let's count how many moons we can see from here. Uranus has 27 moons in total," Dr. H said with a smile.

She and BB pointed out every moon they could find. Some were big, some were small, and some were quite lopsided. It took almost a full day, but they eventually spotted all 27 of Uranus's moons.

27

BB's windows began to fog up from the cold, so they decided it was time to go home. It took BB two tries to start up before he flew off the planet. Dr. H bundled herself up in the driver's seat as they traveled back home. It would be a long drive home because they were close to the last planet in their solar system.

# Discussion Questions

Discussion questions provided by Theresa Apodaca, the 2009 New Mexico Academy of Sciences Science Teacher of the Year. Theresa began a career in teaching after graduating from the University of Tennessee in Knoxville with two science degrees. She has won numerous awards for being an excellent educator. Her teaching career started in Oak Ridge, TN where she taught second grade for two years and won several grants for an outdoor science lab which included parents, students, teachers and community members.

After transferring to New Mexico, she has been recognized several times for being teacher of the year in the school's district. In 2009, Theresa received NM Science Teacher of the Year from the NM Academy of Sciences. She also received Outstanding Educator for Student Achievement from the NM State Board of Education in 2015. Most recently she received an Excellence in Teaching award from the NM Education Department for STEM education. She wants to help her students understand that a person can always do better. Her knowledge and many life experiences have helped her become an excellent educator and community member. Teaching has fulfilled her life.

1. As Dr. H and BB approached Jupiter they noticed that Jupiter spins very. quickly. One day on Jupiter lasts about 10 hours; what would your day be like here on Earth, if there were only 10 hours in one day?

2. Dr. H and BB also noticed colorful bands of gas surrounding the planet. What types of gases make up Jupiter's atmosphere? Why are there different colors of bands surrounding the planet?

3. When they landed on Jupiter Dr. H and BB found it "riddled with storms". What does riddled mean in this sentence? What is the reason for so many strong storms on Jupiter? What kind of storms does Dr. H and BB encounter?

4. Dr. H explained that the gravity is so much stronger on Jupiter and a person would weigh two and a half times heavier. If you were on Jupiter how much would you weigh? Why is the gravity stronger on Jupiter than on Earth?

5. There is a big red spot on Jupiter which has been identified as a storm going on for over three hundred years. How do we know this information about the big red spot?  Why is the spot red?

6. Saturn is surrounded by a spectacular set of rings. BB was surprised that the rings were made up mostly of rocks. What sizes do you think the rocks are; find evidence to support your answer.

7. H explained a year on Saturn is about 30 years on Earth! Knowing this, how long does it take Saturn to orbit the Sun?

8. Dr. H and BB explored Uranus next, and the first thing they noticed was that Uranus is rotating on its side! What explanations have scientists given for this to happen? What theory do you agree with and why

9. Because Uranus is so far away from the Sun it was extremely chilly on the surface for Dr. H. How cold is the atmosphere on Uranus? Compare the temperature to Earth's. Would you be able to adapt to the atmosphere on Uranus?

10. Dr. H and BB discovered Uranus has 27 moons; how many can you name? Which moon is the biggest and which one is the smallest? Compare and contrast how they look

11. It took Dr. H and BB almost one day to count Uranus's moons. How many Earth hours is one day on Uranus?